Y STORI RHIFAU

THE NUMBER STORY

SMALL BOOK ONE

ENGLISH - WELSH

*Numbers Teach Children
Their Number Names*

written and illustrated by

MISS ANNA

Early Reader Edition of *The Number Story 1*
Bronze Medal Winner, 2016 Wishing Shelf Book Award

Library of Congress Control Number: 2018902040

Names: Miss Anna, author.
Title: Number story : numbers teach children their number names / Miss Anna.
Description: Portland, OR: Lumpy Publishing, 2018.
Identifiers: ISBN 978-1-949320-05-3| LCCN 2018902040
Summary: The pictures and rhymes present stories which introduce numbers 0-10.
Subjects: LCSH Numeration—English—Welsh--Pictorial works--Juvenile literature. | BISAC JUVENILE NONFICTION /
Languages: English—Welsh
Classification: LCC QA141.3 .M57 2018 | DDC 513—dc23

Publisher: Lumpy Publishing
Website: www.missannabooks.com
Email: missanna@missannabooks.com

Paperback: ISBN 978-1-949320-05-3
Printed in the U.S.A.   1 3 5 7 9 10 8 6 4 2

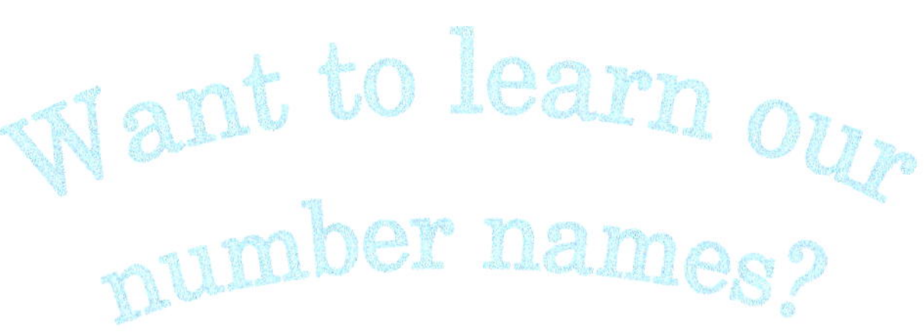

Hoffech chi ddysgu yr enwau o rhifau?

It is very easy and a lot of fun!

Mae'n hawdd iawn ac yn llawer o hwyl!

Say-along our little jingle

Canwch gyda ni ein stori fach!

starting from Number One!

Byddwn yn dechrau o rhif un!

1

ONE   looks like my one finger.

UN

edrych fel fy un bys.

ONE!
UN!

2
TWO trails a tail.
DAU
llwybrau yn gynffon.

A TAIL! Y CYNFFON!

3

THREE has bumps.

TRI

ganddo lympiau.

BUMPY! ANWASTAD!

4

FOUR   carries a sail.

PEDWAR

yn cario malwen.

4
A SAIL!
CWCH HWYLIO!

5

FIVE   is a racing track.

PUMP

yn drac rasio.

VROOM
VROOOM!

SIX curves like a snail.

CHWECH

cromliniau fel malwen.

A SNAIL!　MALWEN!

7
SEVEN has a sharp angle.
SAITH
gyda ongl finiog.

BE CAREFUL! IT'S SHARP!

Byddach yn ofalus! Mae'n finiog!

8

EIGHT is rollercoaster rails.

WYTH

yw rheilffordd coaster rholio.

HWRÊ!
YIPPEE!

NINE   is a bubble on a stick.

NAW

yn swigen ar y ffon.

A BUBBLE!
SWIGEN!

TEN   is an eye of a whale.

DEG

yw un llygad o wŷr.

WINK!
WINC!

HELLO! HELO!

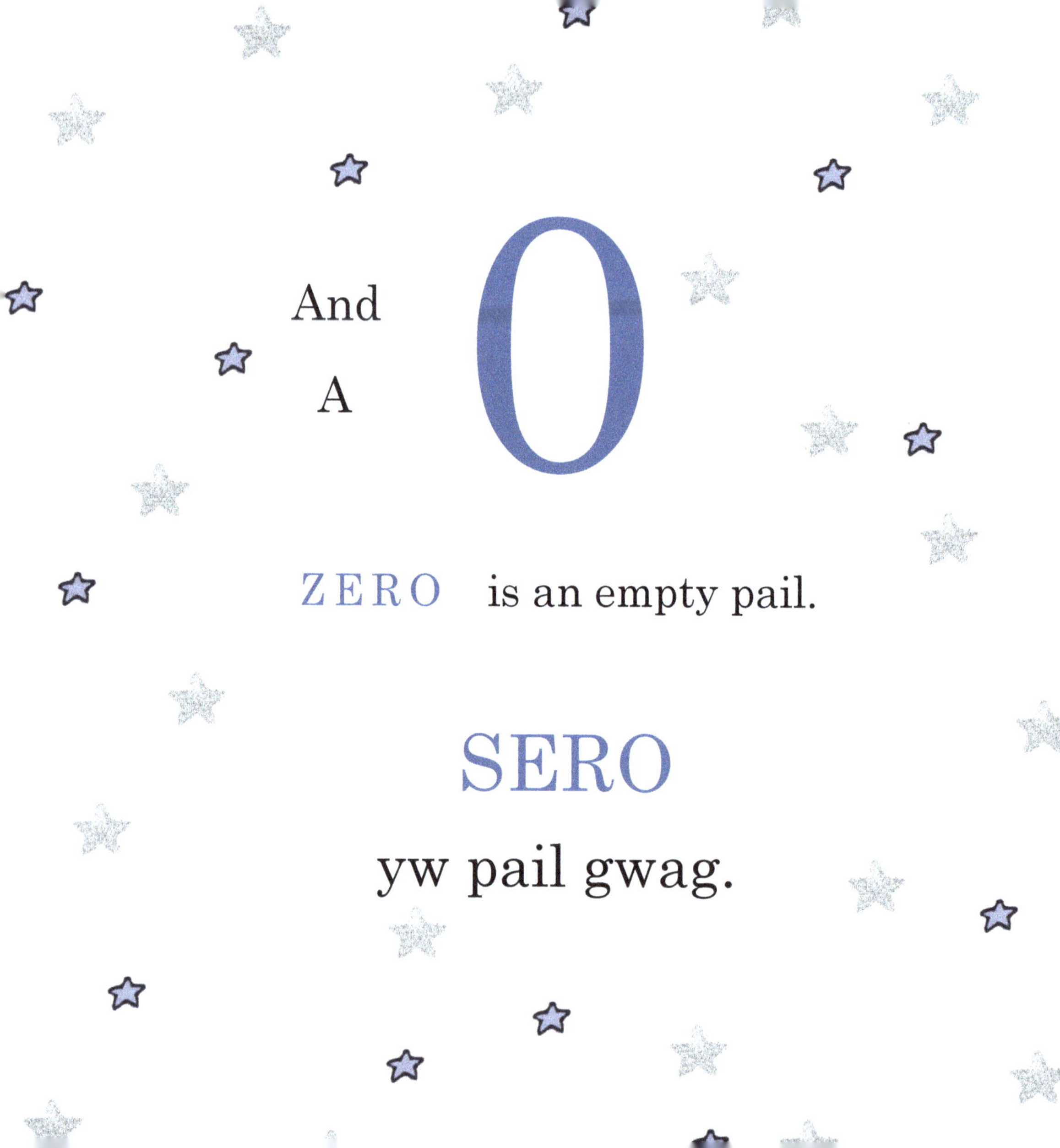

And
A

0

ZERO is an empty pail.

SERO

yw pail gwag.

IT'S EMPTY!
MAE'N GWAG!

Thank you for playing with us today.

We had a lot of fun too!

Diolch am chwarae gyda ni heddiw.

Cawsom lawer o hwyl hefyd!

We are your Number friends,
Zero to Ten,
Who will be here for you~
Ni yw eich ffrindiau rhifau
Sero i Deg.
Byddwn ni yma bob amswer i chi.

Bye-bye now!
See you again soon!
Hwyl nawr!
Welwch chi eto'n fuan!

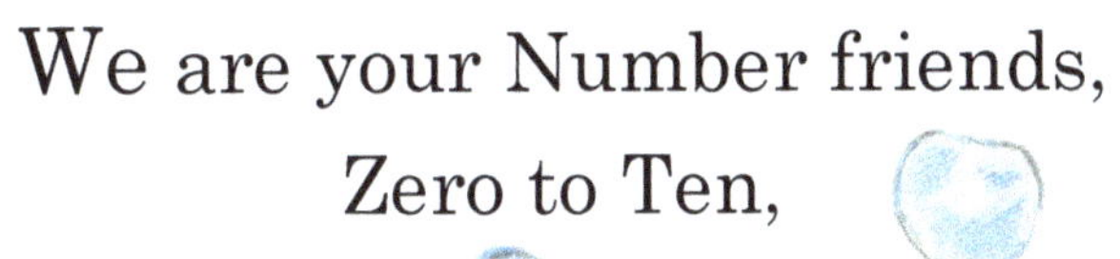

The Numbers are *SINGING* too!

To sing-a-long, look for Miss Anna Number Story
at your favorite music store like iTUNES.

MP3

Numbers 0-10
IDENTIFYING
& COUNTING

Numbers 11-20
& Ordinals

first, second, third...

Numbers 0-100
& Place Values

ones, tens, hundreds...

About Clocks
& Telling Time

hours, minutes, seconds

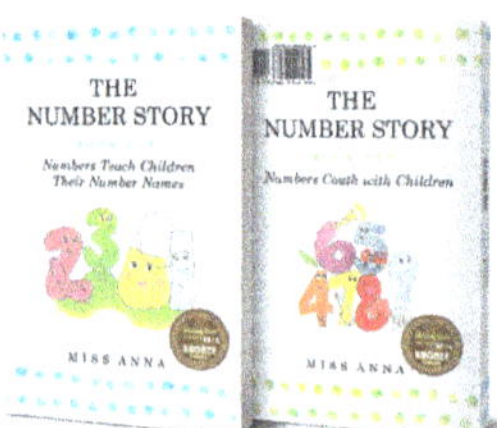

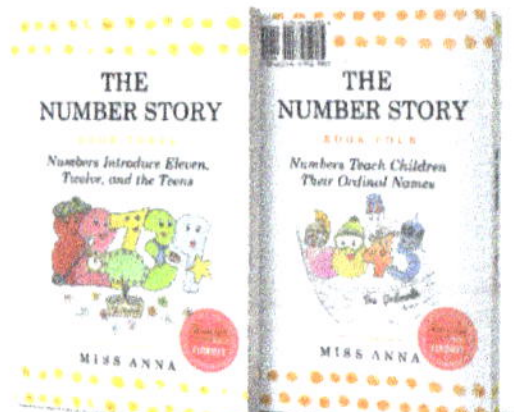

Number Story 1 & 2

isbn: 978-0-996216-48-7

Number Story 3 & 4

isbn: 978-1-945977-01-5

Number Story 5 & 6

isbn: 978-1-945977-06-0

Number Story 7 & 8

isbn: 978-1-949320-40-4

For more Miss Anna books to love,
visit us at

www.missannabooks.com

Numbers are working hard all over the world!
Come Travel the World with Us!